Contents

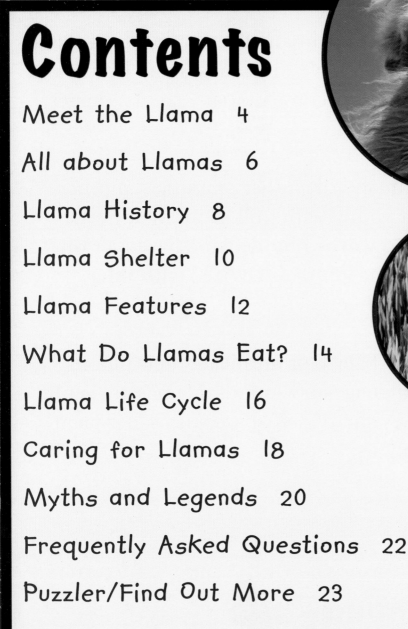

Meet the Llama

Llamas are large farm animals. They have long necks and long legs. Llamas have short tails, small heads, and large, pointy ears. They come in many colors, shapes, and sizes.

Llamas are mammals. Like other mammals, mother llamas feed their young with milk from their body. Mammals are also **warm-blooded** and have hair on their bodies. A llama's body is covered with thick, woolly hairs called fiber.

Llamas are smart. They **communicate** their moods through body language and sounds. Llamas are gentle animals, too. They share their space with other farm animals and humans.

Llamas spit at each other. It is their way of saying "go away" or "leave me alone." Llamas do not often spit at humans.

Farm Animals
Llamas

Heather C. Hudak

Weigl Publishers Inc.

Published by Weigl Publishers Inc.
350 5th Avenue, Suite 3304, PMB 6G
New York, NY 10118-0069
Website: www.weigl.com

Library of Congress Cataloging-in-Publication Data

Hudak, Heather C., 1975-
  Llamas / Heather C. Hudak.
      p. cm. -- (Farm animals)
  Includes index.
  ISBN 1-59036-427-9 (hard cover : alk. paper) -- ISBN 1-59036-434-1 (soft cover :
alk. paper)
  1. Llamas--Juvenile literature. I. Title.
  SF401.L6H83 2006
  636.2'966--dc22
                                              2005034672
Printed in the United States of America
1 2 3 4 5 6 7 8 9 0 10 09 08 07 06

Editor Frances Purslow
Design and Layout Terry Paulhus

Cover: Although llamas are friendly, they do not like to be touched by people
or by other llamas.

Llamas have small, two-toed feet and are very good at climbing mountains.

All about Llamas

Llamas are social animals. They live in groups called herds. Llamas work, too. Some llamas pull carts and pack goods. Other llamas guard sheep.

Llamas are *camelids*. They are part of the camel family. There are three other animals in the camel family. They are **vicunas, guanacos,** and **alpacas**.

There are no breeds, or types, of llamas. Instead, scientists group llamas by their color, hair, body shape, or personality.

Llamas can be divided into groups by their fiber length, too. Llamas can have short, medium, or long hair.

Colors of Llamas

Bay	Black Calico	Brown
• Reddish brown color on body • Black legs • Some black on face	• Mostly black in color • Few red spots near white markings	• No black hairs anywhere on body • Head and legs are often a darker brown

Paint	Seal Bay	White
• Solid color with a splash of white • Sometimes the white splash covers most of the llama's body	• Black or dark brown • Black head and legs • Born black but lightens over time	• White color • Light-colored spots on skin

Llama History

Three million years ago, llama-like animals lived in South America. Llamas were some of the first animals to be tamed. About 4,000 to 5,000 years ago, wild guanacos were bred and tamed in Peru. Llamas come from tamed guanacos.

For thousands of years, **indigenous peoples** in South America have used llamas for food, clothing, fuel, and shelter. Woolen blankets were made from llama fiber. Their meat was used as food. Their droppings were used as fire fuel.

Today, llamas are used mainly as pack animals. They carry goods from one place to another. Adult llamas can carry about 200 pounds (91 kilograms) for up to 12 hours at a time.

There are about 7 million llamas and alpacas in South America. There are about 100,000 llamas in the United States and Canada.

Baby guanacos are called chulengos. When a guanaco gives birth, the chulengo is able to walk immediately.

Llama Shelter

Many llamas live on farms, hills, and mountains. On farms, llamas spend a great deal of time in large, open fields called pastures.

Llamas need shelter to protect them from poor weather. Three-sided shelters work well on pasture. Trees or shelters with a roof provide llamas with shade from the Sun. Llamas may need an air-conditioned area or large fans for very hot weather. In cold places, llamas need a shelter with doors to keep out rain and snow.

Llamas should be kept in a fenced area. Fences must be high. Llamas can jump over low fences. They can also crawl under fences that do not reach to the ground.

Llamas are great work animals and family pets. They are gentle by nature and easy to care for.

Fences keep llamas in and predators out.

Llama Features

Llamas are **sturdy** animals. Their strong muscles help them work and travel long distances.

Llamas are **adapted** to living in the mountains. Most mammals have a difficult time breathing high up in the mountains. This is because the air is thin there. Thin air holds less **oxygen**. Llamas have a special type of blood that helps them breathe easily at high **altitudes**. They are the only mammal that has this trait. Llamas have many other features that make them a **unique** farm animal.

FEET
Llamas have two-toed feet. Each toe has a toenail and pad. The pad helps grip the ground.

EARS
Llamas have big, long ears. They have better hearing than humans.

EYES
Llamas have three eyelids and long lashes. These features help protect their eyes from dust and sand.

FLEECE
Llama hair comes in 22 colors. Farmers collect fiber by brushing or **shearing** a llama.

What Do Llamas Eat?

Llamas are herbivores. They eat plants, such as grass, hay, and herbs. One llama can eat about four bales of hay each month.

Farmers feed llamas vitamins to give them more **nutrients**. Some farmers also add corn, barley, and soybeans to llama feed. Llamas should not eat oats. Oats hurt their stomachs and may cause **ulcers**. Llamas must have fresh water nearby.

Llamas are ruminants. Ruminants have many parts to their stomach. Llamas have three parts to their stomach. To begin eating, llamas chew their food a bit before swallowing it. Then, they spit up the food in their mouth and chew it some more. The chewed food is called cud. Llamas chew the cud until it is completely digested, or broken down.

Fascinating Facts

Llamas eat for six hours and chew cud for eight hours each day.

Llamas eat grasses, hedges, and young tree shoots.

Llama Life Cycle

Mother llamas are called dams. Father llamas are called sires. Crias are newborn llamas.

Dams have one baby at a time. They carry the baby in their belly for about 350 days. Crias are born in spring.

Newborn

At birth, a cria weighs between 18 and 35 pounds (8 and 16 kg).

A cria can stand and walk one hour after birth.

6 Months to 2 Years Old

Between 6 months and 2 years of age, llamas grow quickly. They gain about 1 pound (0.5 kg) of weight each day.

At six months of age, young llamas are called weanlings. This is because they stop drinking milk from their mother. Weanlings begin to eat solid foods, such as hay and grains.

Dams smell and touch their babies with their nose. They talk to their babies by humming. Young llamas stay close to their mothers.

Llamas live for about 20 years. With special care, some llamas can live even longer.

Adult

Llamas are full-grown at 3 years of age. From head to toe, adult llamas stand between 65 and 72 inches (1.6 and 1.8 meters). They weigh between 250 and 400 pounds (113 and 180 kg).

Caring for Llamas

Llamas need special care. They should be taken to a veterinarian, or animal doctor, once a year. Llamas must be **wormed**. They need to get their **vaccinations** as well.

Llama fiber tangles easily and should be softly brushed with gentle tools. A boar-bristle brush is best for grooming llamas.

Llamas should be sheared. In hot places, this is important because it helps keep llamas cool. Shearing is done using special scissors or electric shears.

Llamas' toenails must be trimmed regularly. If they are not trimmed, llamas may become **lame**.

Useful Websites

To learn more about caring for llamas, visit the Roseland Llama Farm at: **www.llamas.co.uk**.

Most llamas are sheared once every two years. Each llama produces about 7 to 8 pounds (3 to 4 kg) of fiber during this time.

Myths and Legends

Llamas are featured in the myths and legends of many South American cultures. They were valued highly in the past. People who owned many llamas were thought to be rich. Llamas were sometimes called "camels of the clouds" or "ships of the Andes." This is because they were used for transportation and fiber by people who lived high in the Andes Mountains.

Llamas carried goods to Machu Picchu, an ancient city in Peru.

Llama Saves the Day

The Huarochiri are indigenous peoples from Peru. This is their tale about how a llama saved all humans.

One day, a man tried to tie his llama to a post. The llama would not let him. The man asked the llama what was wrong. The animal told him that the world would end in five days. The llama said the sea would rise. It would cover all of Earth.

The man asked how he could be saved. The llama told him to climb to the top of a high mountain, called Villca-coto. The man did. He took the llama with him. Many animals were there.

Soon, the sea rose, and Earth was flooded. Five days later, the floods stopped. The only human to live was the man with the llama. All people on Earth came from this man.

Frequently Asked Questions

What basic supplies do I need to care for a llama?

Answer: Some basic supplies include feed pans and dishes, water and hay buckets, heat lamps, a fan, a first-aid kit, a shovel, and a scale to record the llama's weight.

How do I know if my llama is ill?

Answer: It is sometimes hard to tell if a llama is ill. It may not show signs until it is very sick. Signs might include a lack of interest in drinking or eating. A sick llama might have a fever or have trouble breathing.

What kinds of things can be made from llama fiber?

Answer: Llama fiber is made into many items. Hats, scarves, blankets, backpacks, rugs, jackets, ponchos, vests, and wall hangings are some of the items made from their hair.

Puzzler

See if you can answer these questions about llamas.

1. What are baby llamas called?
2. Where did llamas first come from?
3. What can llamas be used for?
4. What do llamas eat?
5. Where do llamas live?

Answers: 1. Crias 2. South America 3. To guard sheep, carry goods, pull carts, provide fiber for making cloth 4. Grass, herbs, plants, hay, corn, soybeans, barley 5. On farms, pasture, hills, mountains

Find Out More

There are many more interesting facts to learn about llamas. If you would like to learn more, take a look at these books.

Bennett, Marty McGee. *The Camelid Companion*. Arcadia, CA: Raccoon Press, 2001.

Palazzo-Craig, Janet. *How Llama Saved the Day: A Story from Peru*. Mahwah, NJ: Troll Publishers, 1999.

Words to Know

adapted: adjusted to the natural environment

alpacas: tame relatives of llamas

altitudes: the heights of places above sea level

communicate: to express feeling or thought

guanacos: wild relatives of llamas

indigenous peoples: the first people to live in a region

lame: having trouble walking

nutrients: parts of food that nourish living things

oxygen: a gas in air

shearing: cutting or trimming off

sturdy: strong

ulcers: sores lining the stomach

unique: special or different

vaccinations: medicines that prevent disease

vicunas: smaller relatives of llamas

warm-blooded: having a constant body temperature

wormed: to remove worms that cause diseases

Index